Henry Davey

A Description of the Differential Expansive Pumping Engine

Henry Davey

A Description of the Differential Expansive Pumping Engine

ISBN/EAN: 9783337337766

Printed in Europe, USA, Canada, Australia, Japan

Cover: Foto ©berggeist007 / pixelio.de

More available books at **www.hansebooks.com**

A DESCRIPTION

OF THE

DIFFERENTIAL EXPANSIVE

PUMPING ENGINE,

GIVING

PRACTICAL EXAMPLES OF ENGINES AT WORK,

AND

A LIST OF SOME OF THE ENGINES ALREADY MADE,

WITH

USEFUL NOTES AND FORMULA FOR PUMPS AND
PUMPING ENGINES.

BY

HENRY DAVEY, M.I.C.E., F.G.S., &c.

HATHORN DAVEY AND CO.,

ENGINEERS, LEEDS,

MAKERS OF

Pumping Machinery of all kinds, Waterworks and Mining Plant,
and Hydraulic Machinery, &c., &c.

PREFACE.

THE success which has attended the introduction of the Differential Engine, and the request I often receive for information respecting it, has induced me to publish a description of the engine with illustrations of the different types which are in use; and in doing so I have appended a few notes and memoranda which may be of some use to practical men engaged in pumping operations.

HENRY DAVEY.

DIFFERENTIAL PUMPING ENGINES.

LIST OF SOME OF THE ENGINES ALREADY MADE, BETWEEN 30 AND 585 H.-P.

No.	Horse-Power.	No. of Gallons Raised per Hour.	Height to which Water is Raised.	No.	Horse-Power.	No. of Gallons Raised per Hour.	Height to which Water is Raised.
1	585	120,000	668	46	112	19,800	600
2	585	120,000	720	47	112	48,000	180
3	476			48	112	48,000	180
4	462			49	112		
5	433			50	106	15,600	855
6	406	120,000	435	51	95	15,000	400
7	394	100,000	600	52	95	15,000	400
8	340	152,174	200	53	92	40,000	240
9	330	84,000	600	54	89		
10	312	42,000	910	55	86	300,000	20
11	312	42,000	910	56	85		
12	304	37,200	200	57	80	72,000	80
13	302			58	78	24,000	300
14	254	72,000	360	59	78	24,000	240
15	254		600	60	78	18,000	480
16	254	60,000	345	61	78		
17	254	30,000	920	62	78		
18	254			63	71·3		
19	254			64	71·3		
20	254			65	71·3	115,800	70
21	235	37,200	1,200	66	71		870
22	235	37,200	1,200	67	70		300
23	230			68	66	39,000	220
24	230			69	64		
25	230			70	62	62,500	190
26	223		1,500	71	62	62,500	190
27	217	46,800	600	71A	60	39,000	240
28	200			71B	60	39,000	240
29	198	24,000	1,100	72	57	13,440	555
30	198	30,000	450	73	51		
31	193			74	50	11,400	350
32	193			75	50	11,400	350
33	193			76	48		
34	193			77	44	12,000	260
35	183	90,000	420	78	44	13,200	262
36	168	60,000	500	79	41	420	750
37	159	60,000	323	80	41	50,000	100
38	154	48,000	600	81	41	12,000	290
39	154	37,600	410	82	41	6,000	480
40	140	24,000	720	83	41		
41	135	7,200	600	84	35	18,000	100
42	125	36,000	390	85	35	6,600	465
43	125			86	34	12,000	480
44	115	54,000	1,220	87	30	30,000	151
45	115	15,000	600				

A DESCRIPTION

OF THE

DIFFERENTIAL EXPANSIVE PUMPING ENGINE.

THE COMPOUND DIFFERENTIAL PUMPING ENGINE.

THE Differential Engine exists in two distinct types, viz.: the single cylinder, and the compound engine; the latter admitting of being worked with high degrees of expansion, is capable of realizing the greatest economy of fuel. The chief peculiarity in the invention, is the simple manner in which the engine is made perfectly safe in working under all conditions of load, automatically varying its supply of steam in proportion as the load on the engine increases or decreases; the distribution of steam being such, that the pumping is performed without shock.

In designing pumping machinery—as also in designing steam machinery of all kinds—the three great questions which should be relatively considered, are economy of fuel, economy of maintenance, and economy of construction.

Economy in fuel must always be a very important consideration; our steam engines consume annually 37,000,000 tons of coal, which at the present moment may perhaps be reckoned at an average of 15s. per ton, representing over 27,000,000l. sterling; an economy of 25 per cent. would therefore effect a saving of nearly 7,000,000l. annually.

With colliery pumping engines it is not unusual to find a consumption of from 12 to 14 lb. of coal per horse-power per hour. A good compound engine will work with less than a quarter of that amount of fuel. The saving to be effected on 400 horse-power of actual work, by the substitution of a compound engine for a non-expansive engine, would be, at the lowest estimate, 36 tons in 24 hours; which taken at 5s. a ton at the pit's mouth would amount to the sum of 2700l. per annum.

As a question of first cost only the necessary boiler power must be taken into account, and it very often occurs that the total cost of engine and boilers is in favour of the most costly and most economical engine.

The leading principle of economy is expansion, and the engine which will work with the greatest amount of expansion is, *cæteris paribus*, the most economical. There are, however, certain conditions necessary to economical working. The resistance to be overcome in pumping, is almost constant, and the force applied to overcome that resistance by the expansion of steam, varies. The condition of the two forces is graphically represented in Fig. 1,

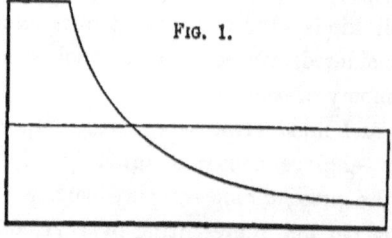

FIG. 1.

where the resistance of the pump is shown by a parallelogram, and the expansive force of steam, by a parabola. It is evident that the mean of the two forces must coincide, but the extremes greatly vary. The steam pressure is too great at the commencement of the stroke, and too small at the end. A means then is required whereby work may be stored up, whilst the piston is moving through the beginning of the stroke, and given out again whilst it is further moving towards the end of the stroke. That function is performed in the Cornish engine by the inertia and momentum of the pit work beam column, and other inert matter; and in the rotative engine, by that of the fly-wheel. It is evident that when a high degree of expansion is employed in a single cylinder an enormous strain (above the resistance of the pump) is put on the engine at the commencement of the stroke; and also that the maximum piston speed must be very great.

These are two of the most serious difficulties surmounted by the Compound Differential Engine. A range of expansion which would produce a variation of strain of six to one in a Cornish engine, would only give two and a half to one in the Compound Engine; that is to say, the strains are nearly three times as great in the Cornish engine.

The importance of thus reducing the strains on the machinery is obvious. The engine may be made lighter, with greater security against breakages. The foundations become cheaper, whilst the speed of the engine is rendered more uniform.

In the Compound Differential Engine, not only are the strains and maximum speeds reduced for a given ratio of expansion, but the *effective* piston speed is increased, because the engine is double instead of single acting.

The following comparison of the two systems—Cornish and Compound Differential—are taken from actual tests in practice.

	Initial Pressure.	Ratio of Expansion.	Average Pressure.	Maximum Piston Velocity per Minute.	Relative Strains on Engine.	Effective Piston Speed.
	lb.		lb.	feet.		
Cornish	31	3	16	600	1·8	100
Ditto	45	4½	19	500	2·26	80
Differential	43	6¼	13	228	1·4	168
Ditto	80	8	24	220	1·37	150

Fig. 2 is a velocity diagram from a Compound Differential Engine with 30-inch and 60-inch cylinders, making twelve strokes per minute, and expanding six times, and Fig. 3 that of a Cornish engine expanding twice.

FIG. 2.

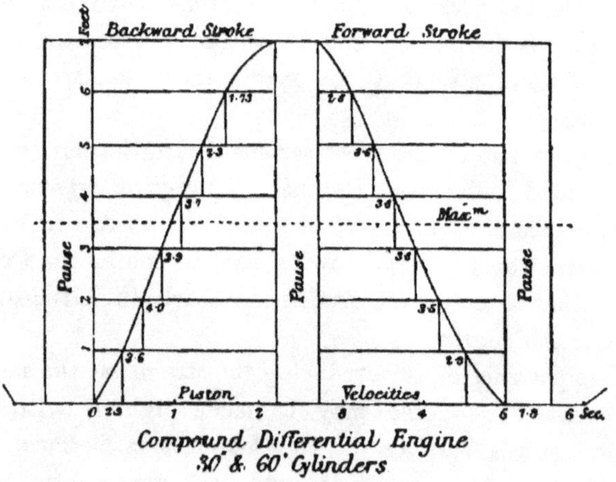

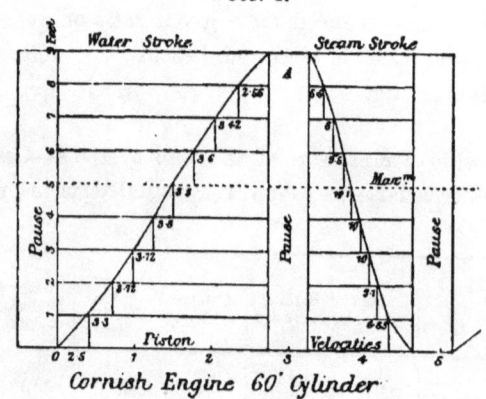

Compound Differential Engine
30' & 60' Cylinders

FIG. 3.

Cornish Engine 60' Cylinder

In the Compound Engine, the initial pressure, ratio of expansion, average pressure and effective piston speed are increased, whilst the strains on the engine and the maximum speed of the piston are reduced.

In practice, the greatest economy is obtained with from six to eight fold expansion, when the initial pressure is about 100 lb.

Fig. 4.

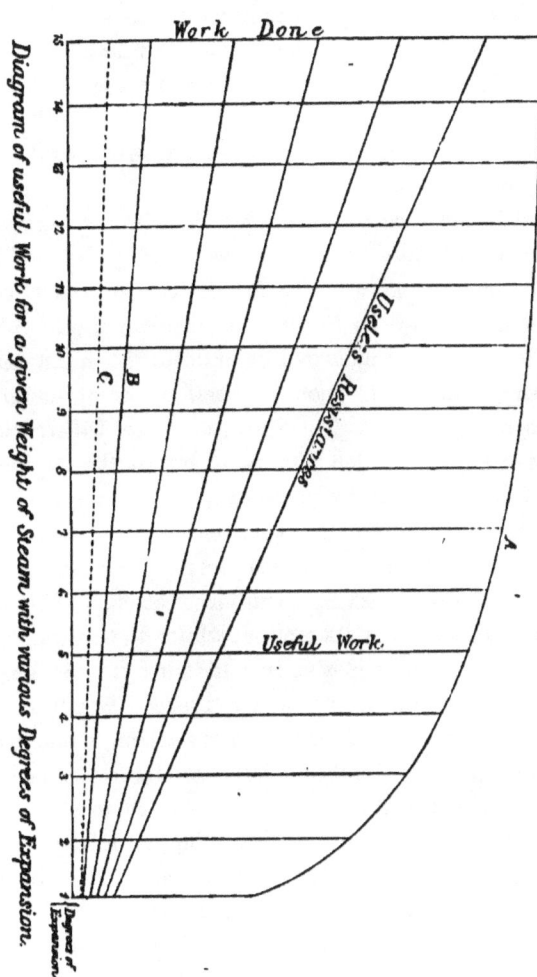

per square inch absolute.   That such a result might be expected on theoretical considerations is seen from an inspection of Fig. 4.

The ordinates of the curve A, represent the work done with

various grades of expansion, from one to fifteen; whilst the straight lines inclined to the base line, cut off the portion of the work lost by useless resistances; each division representing 4 lb. If 2 lb. be taken as the back pressure, and 2 lb. the friction of the engine, and the friction be assumed to increase in the same proportion as the capacity of the cylinder, then the front line B, will cut off from the ordinates the portion of the work lost; the remaining being the useful effect.

It is readily seen how small is the increase of useful effect after expanding eight times.

Next in importance to economy of fuel in pumping engines— or frequently of more importance—are safety in working, and immunity from stoppages. The distribution of steam should be effected in such a way, as to cause no shock or slip in the pumps; and in the event of a sudden loss of load, the engine should be safe; in short, the engine should be self-governing under extreme variations of resistance. To effect this, the Differential Valve Gear was designed, which admits steam to the engine in proportion to the resistance to be overcome; and in case of a sudden total loss of load, reverses the steam to catch the piston. The distribution of steam is effected by coupling the motion of the engine with that of a piston having a uniform velocity. The engine is made to cut off steam by its motion, whilst the uniformly moving subsidiary piston is employed in admitting it. As long as the resistance to the engine is sufficient to prevent its motion becoming relatively equal to that of the subsidiary piston, steam is admitted up to the fixed point of cut off; but should a loss of resistance, or a superior pressure of steam, cause the engine to acquire a speed relatively greater than the speed of the subsidiary piston, then the motion of the steam valve would be reversed earlier, and the supply of steam would be adjusted to the altered conditions. The *modus operandi* is best illustrated by the following diagrams.

The action of the Differential Valve Gear is illustrated in the diagrams Figs. 12 to 15, Plate 9. These diagrams are not drawn to scale, but are intended to show clearly the action of the gear; whilst Fig. 11, Plate 10, shows a practical example of its application to a Compound Engine. The main slide valve G, Fig. 12, is actuated by the piston rod through a lever H, working on a fixed centre, which reduces the motion to the required extent and reverses its direction. The valve spindle is not coupled direct to this lever, but to an intermediate lever L, which is jointed to the first lever H, at one end; the other end M, is jointed to the piston rod of a small subsidiary steam cylinder J, which has a motion independent of the engine cylinder; its slide valve I, being actuated by a third lever N, coupled at one end to the intermediate lever L, and moving on a fixed centre P, at the other end. The motion of the piston in the subsidiary cylinder J, is controlled by a cataract cylinder K, on the same piston rod, by which the motion of this piston is made uniform throughout the stroke; and the regulating plug Q, can be adjusted to give any desired time for the stroke.

The intermediate lever L, Plate 12, has not any fixed centre of motion, its outer end M, being jointed to the piston rod of the subsidiary cylinder J, the main valve G consequently receives a differential motion compounded of the separate motions given to the two ends of the lever L.

If this lever had a fixed centre of motion at the outer end M, the steam would be cut off in the engine cylinder at a constant point in each stroke, on the closing of the slide valve by the motion derived from the engine piston rod; but inasmuch as the centre of motion at the outer end M, of the lever shifts in the opposite direction with the movement of the subsidiary piston J, the position of the cut off point is shifted and depends upon the position of the subsidiary piston at the moment when the slide valve closes. At the beginning of the engine stroke, the subsidiary piston is moving in the same direction as the engine piston, as shown by the arrows in Fig. 12; and in the instance of a light load as illustrated in Fig. 13, the engine piston having less resistance to encounter, moves off at a higher speed, and sooner overtakes the subsidiary piston, moving at a constant speed under the control of the cataract; the closing of the main valve G, is consequently accelerated, causing an earlier cut off. But with a heavy load, as in Fig. 14, the engine piston encountering greater resistance moves off more slowly, and the subsidiary piston has time consequently to advance further in its stroke before it is overtaken, thus retarding the closing of the main valve G, and causing it to cut off later. At the end of the engine stroke Fig. 15, the relative positions become reversed from Fig. 12, in readiness for the commencement of the return stroke.

The subsidiary piston J, Fig. 11, Plate 10, being made to move at a uniform velocity by means of the cataract K, the cut off consequently takes place at the same point in each stroke, so long as the engine continues to work at a uniform speed; but if the speed of the engine becomes changed in consequence of a variation in the load—if, for instance, the load be reduced, causing the engine to make its stroke quicker, the subsidiary piston has not time to advance so far in its stroke before the cut off takes place, and the cut off is therefore effected sooner, as in Fig. 13. On the contrary, if the load be increased, causing the engine stroke to be slower, the additional time allows the subsidiary piston to advance further before the cut off takes place, and the cut off is consequently later, as in Fig. 14.

From the foregoing description of the Valve Gear, it will be understood that every erratic motion of the engine, alters the relative position of the valves with respect to the main piston, and in that way the engine checks itself.

So perfect is the action of this gear, that when properly adjusted, the full load may be thrown suddenly off the engine, without any injury resulting. The effect of a sudden loss of load, is to reverse the action of the valves, and *to throw the steam against the motion of the piston, stopping it before the end of the stroke.* Many instances of this have occurred in practice when a pump rod has broken, a pump valve has failed, or a pipe has burst.

At the Croydon Waterworks the load was suddenly thrown off the engine, when it was running at full speed, by the accidental lifting and tilting over of one of the pump valve seatings; but the engine was instantly checked by the automatic action of the gear, and no more shock was caused to the engine than if the accident had not occurred. Such an accident happening with a Cornish engine would have caused a most serious breakdown.

At the St. Helens Waterworks the mains burst without causing the slightest damage to the Differential Engine; and several such instances have occurred in practice.

THE DIFFERENTIAL VALVE GEAR, AS APPLIED TO
BEAM AND BULL ENGINES, &c.

The illustration on the opposite page represents a perspective view of the gear. It consists of a lever *a*, called the main lever, by means of which motion is given to the valves through a rod *b*. The motion of the engine is given to the outer end of the lever, through the rod *c*, by means of a lever of the first order; the long end of which is attached to the plug-rod or any moving part of the engine, where it gets the motion of the piston on a reduced scale; the other end *d*, deriving its motion from the subsidiary cylinder *e*, and being controlled by means of the cataract *f*. The cylinder has a slide valve which is worked by means of a tappet arm on the rod of the piston of a secondary cylinder; the motion of the secondary piston is also controlled by a secondary cataract. The slide valve is, however, free to move with the motion of the hand lever *g*.

It will be seen that there are two handwheels and a lever attached to the cataracts. The function of the large wheel is to regulate the speed of the engine during the stroke; the small wheel is for regulating the pause between the strokes, whilst the hand lever enables the engineman to hand work the engine. A rocking shaft is employed to give motion to the valves of the engine in the usual way, not shown in the engraving.

The action of the Gear may thus be described. Let the engine be "out doors." The engine end of the main lever will then be in its highest, and the opposite end in its lowest position, the secondary lever being lifted so as to admit steam to the bottom of the secondary cylinder. The engine will pause until the piston of the secondary cylinder shall have travelled to the end of its stroke, and have lifted the valve of the subsidiary cylinder.

The pause will be long or short, according to the regulation of the secondary cataract. The piston of the subsidiary cylinder then

FIG. 5.

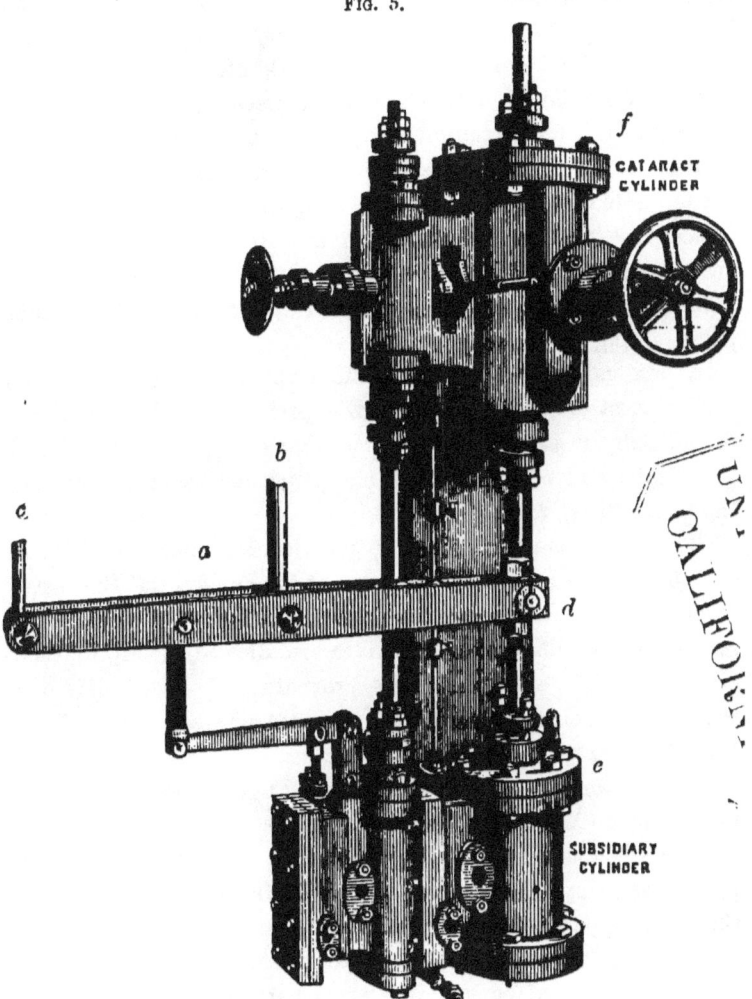

THE DIFFERENTIAL VALVE GEAR AS APPLIED TO BEAM ENGINES.
(Davey's Patent.)

having steam on its lower surface will travel upwards, actuating the main lever with a speed dependent on the adjustment of the cataract. In doing so, the steam valve of the engine will be opened through the medium of the rod *b*, and a rocking shaft, &c., and will be opened quickly because the engine end of the lever, is, for the time being, stationary. Steam now being admitted on the engine piston, it will, after overcoming the inertia of the load, move off at an increasing speed, which is communicated to the engine end of the main lever. The result is, that as the opposite end of the lever is moving uniformly in the opposite direction at the same time, the motion of the centre is soon reversed in direction, and the steam valve which was opened by the motion of the subsidiary piston, is closed by the differential motion, brought into action by the motion of the main piston acting on the same point.

When valves fail to close before the return stroke of the engine, most dangerous shocks are experienced. The shocks act prejudicially in injuring the valves, shortening the durability of the engine and rendering it liable to accident, such as bursting the pump, pipes, &c.

To prevent such shocks is of great consequence, especially in high lifts. The motion of the piston of the Differential Expansive Engine is such, that a pause is produced at the completion of each stroke, during which time the valves are allowed to fall, by their own weight alone, to their seats, preventing the possibility of a shock from sudden closing under pressure, and also preventing a loss of efficiency from slip.

The advantages of the Differential Engine may therefore be briefly summed up as follows:—

1. Economy of fuel, owing to the fact that steam can be expanded as many times as required, while the maximum strain on the working parts is greatly reduced.

2. Safety in the event of a sudden loss of load.

3. Simplicity of parts, few wearing parts, and small frictional resistances, and the foundations are of a very plain and inexpensive character.

The engine can be easily and rapidly moved from one place to another, and erected in a short space of time.

The following are results of a trial with a Compound Differential Engine having 34″ and 64″ cylinders:—

Mean stroke of engine		7 ft. 2¼ in.
„ number of double strokes per minute		8.
„ piston speed in feet per minute		115 ft.
„ pressure of steam in engine room		62 lb.
„ vacuum in condenser		11¾ „

Mean effective pressures in cylinders:

Front of high-pressure cylinder	44·79 lb.
Back „ „ „	48·52 „
Front „ low-pressure „	9·03 „
Back „ „ „	10·37 „

Indicated horse-powers developed:

In high-pressure cylinder	146·42 H.-P.
„ low-pressure „	108·05 „
Total	254·47 „

Condensing Water:

Quantity of water discharged per minute over tumbling bay .. .. .. —	1877 lb.
Mean temperature of do.	103·5°
„ „ injection water	55°
Rise of temperature of the water in the condenser	48·5°
Pounds-degrees of heat discharged from the condenser per minute	91,034 units.
Do. per H.-P.	357·7 „
Consumption of steam per H.-P. per hour	22·0 lb.
Equivalent duty per 112 lb. of coal, assuming each lb. of coal to evaporate 9 lb. of water	90,720,000.
Equivalent duty per 112 lb. of coal, assuming each lb. of coal to evaporate 10 lb. of water	100,800,000.
Effective work performed in pumps	198·6 H.-P.
„ „ in percentage of ind. H.-P.	78·05 %

USEFUL NOTES.

GRAPHIC ILLUSTRATION OF THE DISTRIBUTION OF HEAT IN OUR BEST STEAM ENGINES.

The mechanical equivalent of one unit of heat = 772 foot lb. One unit of heat is that required to raise 1 lb. of water from 32° to 33° Fahr.

One lb. of Welsh coal in a good Cornish boiler will evaporate 10 lb. of water from 212°. The total heat of combustion of the

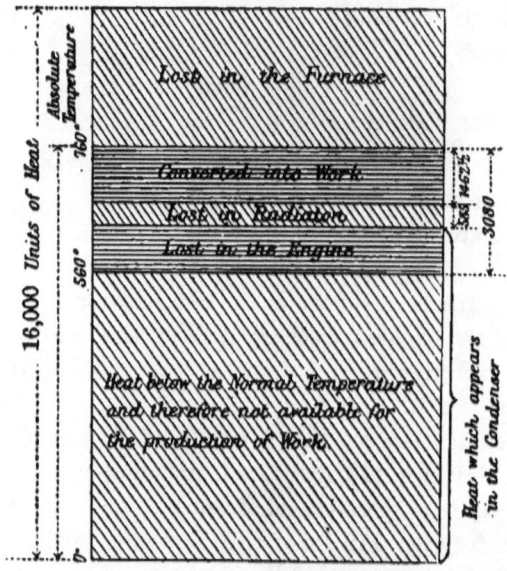

1 lb. of coal is 16,000 units, as shown in the diagram, but the total amount of heat in the 10 lb. of steam produced is only 11,700 units. The difference of the two quantities is that lost in incomplete combustion, in radiation, and in the waste heat of the chimney.

A very economical steam engine uses 10 lb. of steam per half indicated horse-power per hour, or equal to 990,000 units of work. The mechanical equivalent of the 11,700 units of heat is 11,700 × 772, which equals 9,032,400.

Dividing 990,000 by 9,032,400, we get 0·1096 as the portion of mechanical energy which is utilized by conversion of heat into work. That quantity is represented in the diagram. The heat represented by the remaining portion goes away in radiation and in the exhaust steam or condensing water.

TEMPERATURE.

Temperature is a measure of intensity of heat, and to have an absolute value must be reckoned on a scale whose zero indicates no heat.

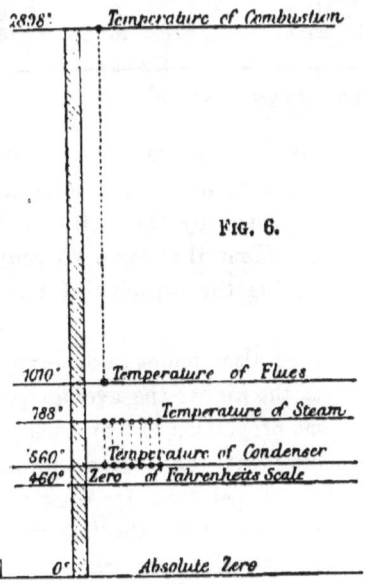

2898° Temperature of Combustion

Fig. 6.

1010° Temperature of Flues
788° Temperature of Steam
560° Temperature of Condenser
460° Zero of Fahrenheits Scale

0° Absolute Zero

When Fahrenheit constructed his scale, which is found on most English thermometers, the absolute zero was unknown. It has since been ascertained, and temperatures reckoned from it are called absolute temperatures, *vide* Fig. 6.

c 2

Let T = highest absolute temperature.

t = lowest   „   „

The maximum theoretical efficiency of any heat engine is
$\dfrac{T - t}{T}$.

TABLE SHOWING A COMPARISON OF THE USEFUL WORK DONE BY EQUAL WEIGHTS
OF STEAM WITH DIFFERENT DEGREES OF EXPANSION AND DIFFERENT BACK
PRESSURES.

NON-CONDENSING ENGINES, STEAM PRESSURE 60 LB. PER SQUARE INCH.				CONDENSING ENGINES, STEAM PRESSURE 60 LB. PER SQUARE INCH.			
Grades of Expansion.	Useful Work.	Back Pressure.		Grades of Expansion.	Useful Work.	Back Pressure.	
a.	*g.*	*u.*	*b.*	*g.*	*u.*	*b.*	*a.*
50·7	2	69·4	16	4	134·8	2	35·7
35·7	4	78·8	„	8	168·8	2	23·1
23·1	8	56·8	„	16	193·6	2	14·1
50·7	2	61·4	20	4	126·8	4	35·7
35·7	4	62·8	„	8	152·8	4	23·1
23·1	8	24·8	„	16	161·6	4	14·1

a = average pressure per square inch. $u = g\,(a - b)$.

NOTE.—The gain as shown above for condensing engines between eightfold and sixteenfold expansion is so small that it is more than covered by the loss caused by the extra cooling effect of the larger engine. It is evident that for a sixteenfold expansion the engine must have double the capacity of that for an eightfold expansion.

The above reasoning also applies to non-expansion engines, for it is self-evident that the higher the average pressure throughout the stroke the less the proportion lost by back pressure.

An engine working with an average pressure of 10 lb. per square inch will lose 20 per cent. by back pressure, whereas if the average pressure were 20 lb., the loss would only be 10 per cent. With practical men it is too common a notion that low-pressure steam is as economical as high pressure for the same ratio of expansion, but, as we have already seen, there is not a greater fallacy.

HORSE-POWER.

An actual horse-power equals 33,000 lb. raised 1 foot high per minute, or 33,000 units of work, which I shall denote by the letters *H.-P.* The term nominal horse-power,—written *N. H.-P.*, —has a varied signification according to the fancies of different engineers, and is therefore made very confusing. It is a term which might be dispensed with, and the actual H.-P. universally adopted as the standard, to advantage.

The Admiralty rule for *N. H.-P.* is—

$$N.\ H.\text{-}P. = \frac{7\,A\,V}{33,000} \text{ or } \frac{D^2\,V}{600}$$

V = velocity of piston in feet per minute.
A = area of cylinder.
D = diameter of cylinder.
S = stroke of engine in feet.

Ordinary rules—

$$N.\ H.\text{-}P. = \frac{D^2\,\sqrt[3]{S}}{15 \cdot 6} \text{ for high-pressure engines.}$$

$$N.\ H.\text{-}P. = \frac{D^2\,\sqrt[3]{S}}{47} \text{ for condensing engines.}$$

For ordinary horizontal high-pressure non-condensing engines of commerce, an approximate rule is $N.\ H.\text{-}P. = \frac{D^2}{12}$, and $H.\text{-}P. = \frac{D^2}{6}$, which is approximately right for an initial pressure of 50 lb. and cut off at half stroke, with 300 feet piston speed per minute.

PRACTICAL NOTES AND FORMÚLÆ FOR PUMPS AND PUMPING ENGINES.

To find the quantity of water delivered from a given pump or pipe—

Let d = the diameter in inches, then :—

$\dfrac{d^2}{30}$ = the quantity delivered per foot, stroke, or flow in gallons, a little under the theoretical quantity.

A good working velocity of flow for water in the pipes is 200 feet per minute, and the speed of pump piston or ram most suitable for the "Differential Expansive" Pumping Engine, is equal to $\sqrt{L} \times 60$, where L = the length of the stroke in feet.

d = diameter of pipe in inches for 200 feet flow per minute.

p = diameter of plunger in inches—double-acting.

q = number of gallons delivered per minute.

$q = d^2 \times 6\cdot66 = 200\,\dfrac{d^2}{30}$, for a velocity of 200 feet per minute.

n = number of gallons delivered per hour.

$n = d^2 \times 400$.

$q = \dfrac{p^2}{30}\sqrt{L} \times 60$.

q = number of gallons of water raised per minute.

H = the height to which it is raised in feet.

$H.\text{-}P.$ = effective horse-power of engine without friction.

$H.\text{-}P. = qH \times 9\cdot000303 = \dfrac{qH}{3,300}$.

h = head of water in feet.

p = pressure in lb. per square inch.

f = pressure in lb. per square foot.

$p = h \times 0\cdot433$.

$h = p \times 2\cdot31$.

$f = h \times 62\cdot4$.

1 cubic foot of water = $6\cdot24$ gallons = $62\cdot4$ lb. = $\cdot557$ cwt. = $\cdot028$ tons = $6\frac{1}{4}$ gallons approximately.

1 gallon = 10 lb. or $0\cdot16$ cubic foot.

1 cwt. of water = $1\cdot8$ cubic foot = $11\cdot2$ gallons.

1 ton of water = $35\cdot9$ cubic feet = 224 gallons.

Weight of sea water = weight of fresh water $\times 1\cdot028$.

A cylinder of spring water 1 inch diameter and 1 fathom long = $2\cdot045$ lb.

DUTY OF PUMPING ENGINES.

A cylinder of spring water, 1 inch diameter and 1 fathom long, weighs 2·045 lb.

L = length of stroke.
N = number of strokes in one month.
H = height of lift in fathoms.
d = diameter of pumps.
q = number of bushels of coal consumed in one month.
$Duty$ = number of lb. lifted 1 foot high per bushel of coal.

$$Duty = \frac{[(H \times (d^2) \times 2·045) \times N \times L]}{q}.$$

The calculation was formerly made for bushels of coal, each weighing 94 lb., but it is now usual to use the cwt, in the place of the bushel.

The usual mode of expressing the efficiency of a steam engine is in terms of coal burnt per horse-power or units of work per hour, but as the efficiency of the boiler is not in such a case distinguished from that of the engine, it is important that a rule should be established by which the efficiency of either might be readily ascertained.

The efficiency of the boiler may be expressed in lb. of water evaporated per lb. of fuel, whilst the efficiency of the engine should be denoted in *units of work done per 1 lb. of steam used.*

A formula for that purpose may be thus constructed :—

Let 28 = number of cubic inches of water in 1 lb. of steam.

I = initial pressure of steam.
S = its specific volume.
R = ratio of expansion employed.

a = average pressure per square inch = $\dfrac{(1 + Hyp.\ Log.\ R)\,I}{R}$.

U = units of work done by 1 lb. of steam,

then $U = a\dfrac{(28 \times S \times R)}{12}$.

Let $I = 100$ lb. per square inch, then :

$S = 270$. See table, page 32.

The following table shows, in units of work, the comparative values of different degrees of expansion with a constant initial pressure equal to 100 lb. per square inch; worked out by means of the above formula, which for 100 lb. initial pressure stands thus : $U = a (630 \times R)$.

Initial Pressure.	Ratio of Expansion = R.	Units of Work.	Average Pressure $= \frac{(1 + Hyp.\ Log.\ R)\ I}{R}$.
lb.			lb.
100	0	63000·	100
100	1·25	77017·50	97·8
100	1·66	94958·64	90·8
100	2·00	106596·	84·6
100	3·00	132111·	69·9
100	4·00	150192·	59·6
100	5·00	164115·	52·1
100	8·00	193536·	38·4
100	10·00	207900·	33·0

The formula for the duty of pumping engines * would then become $Duty = \dfrac{(H \times (d^2) \times 2\cdot045) \times N \times L}{W}$, in which $W =$ weight of water in lb., and $duty =$ units of work per lb. of steam.

For the convenience of readily and accurately recording the duty of pumping engines, the water should be measured into the boiler by a meter made to register in lb., and if at the same time the weight of fuel burnt were noted, the *duty* of both engine and boiler could be ascertained. The *duty* of the boiler would be found by dividing the number of lb. of water evaporated by the number of lb. of standard fuel burnt, and the results obtained should then be expressed thus :

Duty of engine = number of units of work per lb. of water evaporated.

Duty of boiler = number of lb. of water evaporated per lb. of coal burnt.

* See page 23.

WATERWORKS.

CONSUMPTION OF WATER IN TOWNS.

15 to 20 gallons per head of population per day in non-manufacturing towns.

20 to 30 gallons per head in manufacturing towns.

Maximum demand 2½ times the average.

Impounding reservoirs should contain 120 days' supply in the rainy districts and 200 days' supply in the less rainy districts of England.

Service reservoirs should contain 3 days' supply.

RAINFALL.

Greatest rainfall in England in 24 hours about 3 inches.

Annual rainfall in England from 20 to 70 inches.

Mean   „   „   „   42 inches.

Mean daily evaporation in England, ·08.

Infiltration in England in Winter,   33 per cent.

„   „   Spring,   35   „

„   „   Summer, 52   „

„   „   Autumn, 48   „

Average for the year, 42   „

WELLS IN VARIOUS GEOLOGICAL FORMATIONS.

In the New Red Sandstone.—Robert Stephenson found when experimenting with wells in this formation around Liverpool, that abundance of water is stored up in the new red sandstone, but the sandstone is generally very pervious, admitting of deep wells drawing their supply from distances exceeding one mile; the permeability is however occasionally interfered with by faults or fissures filled with argillaceous matter. He also reported that there was no possibility of obtaining permanently more than about 1,000,000 or 1,200,000 gallons per day from any one well, and this only when not interfered with by other wells.

Through Chalk into Lower Greensand.—A remarkable artesian well has been bored at Passy in this formation, for supplying water to Paris. The first water-bearing strata was reached at a depth of 1894 feet, but the water did not rise to the surface. At length a true artesian spring was tapped at a depth of 1923 feet, yielding 5,582,000 gallons per day.

Through Tertiaries into Upper Greensand and Chalk.—Wells have been bored into this formation with varying success. At Brighton, a boring is said to yield 1,000,000 gallons per day. At Worthing, 30 feet above the sea and 360 feet deep in the chalk, a well gives a good supply. In the valley of the Lee, wells have been successful up to 100,000 gallons per day near the surface. At Southampton, a depth of 1317 feet was reached without success, and the scheme was abandoned; several other similar undertakings met with a like result.

Domestic Wells generally catch the adjacent percolating water from the surface, and are very liable to receive impurities, unless great care is exercised in selecting the site and ensuring a proper construction.

Dimensions and Weight of Pump Pipes for Mines.

Length of Pipe, 9 Feet.

Diameter of Pipe in inches.	Thickness in inches.	Number of Holes in Flange.	Size of Holes.	Distance between Holes.	Weight. Cwt.	Qrs.	Lb.	Diameter of Pipe in inches.	Thickness in inches.	Number of Holes in Flange.	Size of Holes.	Distance between Holes.	Weight. Cwt.	Qrs.	Lb.
4	¾	4	1⅛	8½	3	0	0	15	1¼	8	1½	20	18	3	0
4½	¾	4	1⅛	9	3	1	14	16	1¼	8	1½	21	19	0	0
5	⅞	6	1⅛	9½	4	1	14	16	1⅜	8	1½	21	22	0	0
6	1	6	1¼	10½	6	0	0	17	1¼	8	1½	22	21	1	0
7	1	6	1¼	11½	7	0	0	17	1⅜	8	1½	22	23	1	14
8	1	6	1¼	12½	8	0	0	18	1¼	10	1½	23	22	2	0
9	1	6	1⅜	14	9	0	0	18	1⅜	10	1½	23	24	3	0
9	1⅛	6	1⅜	14	10	0	0	19	1¼	10	1½	24	23	3	0
10	1	6	1⅜	15	10	0	0	19	1⅜	10	1½	24	26	0	14
10	1⅛	6	1⅜	15	11	1	0	20	1¼	10	1½	25	25	0	0
11	1	6	1⅜	16	11	0	0	20	1½	10	1½	25	30	0	0
11	1⅛	6	1⅜	16	12	1	14	21	1¼	10	1½	26	26	1	0
12	1⅛	6	1⅜	17	13	2	0	21	1½	10	1½	26	31	2	0
12	1¼	6	1⅜	17	15	0	0	22	1¼	10	1½	27	27	2	0
13	1⅛	8	1⅜	18	14	2	14	22	1½	10	1½	27	33	0	0
13	1¼	8	1⅜	18	16	1	0	23	1¼	12	1½	28	28	0	0
14	1⅛	8	1⅜	19	15	3	0	23	1½	12	1½	28	34	2	0
14	1¼	8	1⅜	19	17	2	0	24	1¼	12	1½	29	30	0	0
15	1⅛	8	1½	20	16	3	14	24	1½	12	1½	29	36	0	0

DISCHARGE OF WATER THROUGH PIPES.

180 Feet per Minute.			200 Feet per Minute.			220 Feet per Minute.		
Diameter of Pipe.	Head required per 100 feet.	Gallons per Minute.	Diameter of Pipe.	Head required per 100 feet.	Gallons per Minute.	Diameter of Pipe.	Head required per 100 feet.	Gallons per Minute.
1	4·32	6·110	1	5·33	6·796	1	6·45	7·482
2	2·16	24·441	2	2·67	27·184	2	3·22	29·928
3	1·44	55·117	3	1·78	61·227	3	2·15	67·338
4	1·08	97·888	4	1·33	108·738	4	1·61	119·587
5	0·86	152·944	5	1·07	169·966	5	1·29	186·987
6	0·72	220·282	6	0·89	244·786	6	1·08	269·289
7	0·62	299·903	7	0·76	333·198	7	0·92	366·493
8	0·54	391·682	8	0·67	435·203	8	0·81	478·723
9	0·48	495·807	9	0·59	550·924	9	0·72	606·042
10	0·43	612·089	10	0·53	680·113	10	0·64	748·137
11	0·39	740·718	11	0·49	823·020	11	0·59	905·322
12	0·36	881·629	12	0·45	979·518	12	0·54	1077·408
13	0·33	1034·386	13	0·41	1149·110	13	0·50	1264·458
14	0·31	1199·614	14	0·38	1333·043	14	0·46	1406·472
15	0·29	1377·311	15	0·36	1530·069	15	0·43	1683·450
16	0·27	1566·855	16	0·33	1740·812	16	0·40	1915·392
17	0·26	1769·493	17	0·31	1965·895	17	0·38	2102·991
18	0·24	1983·333	18	0·30	2203·449	18	0·36	2424·168
20	0·22	2448·484	20	0·27	2720·320	20	0·32	2992·176
22	0·20	2962·872	22	0·24	3292·808	22	0·29	3620·664
24	0·18	3525·892	24	0·22	3917·450	24	0·23	4309·008

WATER WHEELS.

Let it be required to find the effective thrust of the crank of an overshot water wheel employed for pumping.

c = length of crank.

r = radius of wheel.

$\dfrac{r - c}{3}$ = effective leverage.

n = number of buckets.

W = weight of water in each bucket.

$$f = \text{friction} = \frac{\left(\dfrac{n}{3} \times W\right) \times \dfrac{r - c}{3}}{5}.$$

p = effective thrust.

$$p = \left[\left(\frac{n}{3} \times W\right) \times \frac{r - c}{3}\right] - f.$$

DISCHARGE OF WATER OVER SILLS.

d = depth from surface of water where at rest, to the top of sill in inches.

c = cubic feet of water discharged per minute for every foot in width of sill.

$c = 5 \cdot 15 \sqrt{d^3}$.

$H = \dfrac{d}{12}$ = depth in feet.

$c = 214 \sqrt{H^3 + \cdot 035\, V^2\, H}$, if the water approaches the sill with a velocity $= V$.

MEASURING WATER PUMPED FROM MINES, &c.

It is very useful to be able to ascertain at any time the quantity of water which is being pumped. A convenient mode of doing it is to lead the water from the pump to a small pond in which the velocity of the stream may be rendered almost *nil*, and to allow the water from the pond to fall over a weir 18 inches wide with vertical sides. A vertical post should be driven in the pond, back from the weir or sill, in still water, and on the post should be painted a gauge in inches and fractions, with zero at the level of the sill. The following table will then show the quantity of water passing over the weir :—

WEIR TABLE FOR 18-INCH OVERFALL.

Depth of Water.	Cubic Feet per Minute.	Depth of Water.	Cubic Feet per Minute.	Depth of Water.	Cubic Feet per Minute.	Depth of Water.	Cubic Feet per Minute.
inches.		inches.		inches.		inches.	
1	7·725	7	142·974	13	361·296	19	639·216
1¼	10·730	7¼	150·637	13¼	372·583	19¼	652·445
1½	14·127	7½	158·55	13½	383·314	19½	665·122
1¾	17·833	7¾	167·04	13¾	393·866	19¾	678·030
2	21·770	8	174·627	14	404·296	20	690·168
2¼	26·016	8¼	183·	14¼	415·550	20¼	703·940
2½	29·490	8½	190·355	14½	426·495	20½	716·790
2¾	35·203	8¾	199·92	14¾	437·605	20¾	730·174
3	40·068	9	208·440	15	447·76	21	742·664
3¼	45·162	9¼	217·23	15¼	460·04	21¼	756·725
3½	50·488	9½	226·185	15½	471·375	21½	770·10
3¾	56·047	9¾	235·149	15¾	482·936	21¾	783·392
4	61·760	10	244·106	16	494·08	22	794·933
4¼	67·627	10¼	253·5	16¼	506·033	22¼	810·738
4½	73·648	10½	262·8	16½	517·729	22½	824·415
4¾	79·902	10¾	272·28	16¾	529·564	22¾	832·239
5	86·309	11	281·625	17	546·576	23	851·516
5¼	92·794	11¼	292·11	17¼	553·457	23¼	866·049
5½	99·610	11½	301·20	17½	565·470	23½	879·795
5¾	106·381	11¾	311·47	17¾	577·690	23¾	893·991
6	113·406	12	321·06	18	589·036	24	907·100
6¼	120·586	12¼	331·2	18¼	602·271	24¼	922·519
6½	127·920	12½	341·367	18½	614·601	24½	933·647
6¾	135·331	12¾	351·69	18¾	627·185	24¾	951·179

PIPE JOINTING.

Tho weight of lead required for jointing socket pipes may bo approximately estimated by reckoning 1 lb. per inch diameter of pipe for each joint up to 8 inches diameter; 1¼ lb. from 8 inches to 16 inches diameter, and 1½ lb. from 16 inches to 24 inches diameter. In the small pipe joints the thickness of lead should be about $\frac{5}{16}$ inch, and in the large ones about $\frac{7}{16}$ inch. The depth of socket is usually about 4½ inches, the pipe measuring 9 feet 4½ inches over all.

Strength of cylinders—circumferential strain:

r = internal radius.

p = pressure per square inch.

c = cohesive strength of the metal.

x = the required thickness.

$$x = \frac{p \times r}{c - p}.$$

PROPORTIONS OF CAST-IRON FLANGE PIPES.

Diameter of Pipe.	Diameter of Flange.	Thickness of Flange.	Number of Bolts.	Diameter of Bolts.	Diameter of Circle of Bolts
inches.	inches.	inch.		inch.	inches.
1½	4½	½	3	⅜	3¼
2	5¼	½	3	$\frac{7}{16}$	3¾
2½	6	⅝	4	$\frac{7}{16}$	4½
3	6½	⅝	4	½	5
4	8	⅝	4	$\frac{9}{16}$	6¼
5	9¼	¾	4	$\frac{9}{16}$	7½
6	10½	¾	6	⅝	8¾
7	12	¾	6	⅝	10
8	13¼	⅞	6	⅝	11¼
9	14½	⅞	6	⅝	12¼
10	16	1	6	¾	13¼
12	18¼	1	6	¾	16

WEIGHTS AND PROPORTION OF CAST-IRON SOCKET PIPES, FOR A HEAD
OF ABOUT 200 FEET.

Diameter of Pipe.	Length without Socket.	Depth of Socket.	Lead Joint.			Average Weight of the Pipe.			Average Weight of Quarter Bends.		
			Thickness. inch.	Depth. inches.	Weight. lb.	cwt.	qrs.	lb.	cwt.	qrs.	lb.
Inches.	feet.	Inches.									
$1\frac{1}{2}$	6	3	$\frac{1}{4}$	$1\frac{1}{4}$	1·2	0	1	14	0	1	0
2	6	3	$\frac{1}{4}$	$1\frac{1}{2}$	1·4	0	2	0	0	1	2
$2\frac{1}{2}$	6	$3\frac{1}{4}$	$\frac{1}{4}$	$1\frac{3}{4}$	1·6	0	2	11	0	1	9
3	9	$3\frac{1}{2}$	$\frac{1}{4}$	$1\frac{3}{4}$	2·3	1	0	8	0	1	17
4	9	4	$\frac{5}{16}$	2	4·0	1	2	0	0	2	19
5	9	4	$\frac{5}{16}$	2	5·0	2	0	0	0	3	0
6	9	$4\frac{1}{4}$	$\frac{5}{16}$	$2\frac{1}{4}$	6·5	2	2	0	0	3	3
7	9	$4\frac{1}{2}$	$\frac{5}{16}$	$2\frac{1}{4}$	7·7	3	0	12	1	0	25
8	9	$4\frac{1}{2}$	$\frac{5}{16}$	$2\frac{1}{4}$	8·2	3	2	24	1	2	2
9	9	$4\frac{1}{2}$	$\frac{7}{16}$	$2\frac{1}{4}$	10·4	4	1	0	1	2	18
10	9	$4\frac{1}{2}$	$\frac{8}{16}$	$2\frac{1}{2}$	11·5	4	3	14	3	3	0
12	9	$4\frac{1}{2}$	$\frac{9}{16}$	$2\frac{3}{4}$	15·0	6	2	0	3	3	21

Weight of Pipes.

D = outside diameter in inches.

d = inside diameter.

W = weight of lineal foot in lb.

$W = k\,(D^2 - d^2)$.

$k = 2\cdot45$ for cast iron.

$= 2\cdot64$ for wrought iron.

$= 2\cdot82$ for brass.

$= 3\cdot03$ for copper.

$= 3\cdot86$ for lead.

HYDRAULIC FORMULÆ.

g = force of gravity = $32\cdot2$.

H = head of water in feet.

P = pressure per square inch.

v = theoretical velocity in feet per second.

$H = P \times 2\cdot307$.　　　$P = H \times \cdot4335$.

$\sqrt{2g} = 8\cdot025$.　　　$V = 8\cdot025\,\sqrt{H}$.

$H = \cdot0155\,v^2$.

DELIVERY OF WATER IN PIPES.

D = diameter of pipe in inches.

H = head of water in feet.

L = length of pipe in feet.

W = cubic feet of water discharged per minute.

$$W = 4\cdot72\sqrt{\dfrac{D^5\,H}{L}}$$

$$D = 538\sqrt[5]{\dfrac{L\,W^2}{H}}$$

SATURATED STEAM.			HYPERBOLIC LOGARITHMS.		
Absolute Pressure in lb. per square inch.	Temperature of Steam — Fahr.	Specific Volume.	Ratio of Expansion = R.	Hyp. Log.	Portion of Stroke at which Steam is cut off.
5	162·3	4527	1·25	·223	$\frac{8}{10}$
10	193·3	2358	1·43	·358	$\frac{7}{10}$
*14·7	212·0	1642	1·66	·506	$\frac{6}{10}$
20	228·0	1229	2·00	·693	$\frac{1}{2}$
23	235·5	1075	2·50	·916	$\frac{4}{10}$
30	250·4	838	3·33	1·202	$\frac{3}{10}$
35	259·3	726	4·00	1·386	$\frac{1}{4}$
40	267·3	640	5·00	1·609	$\frac{1}{5}$
45	274·4	572	6·00	1·791	$\frac{1}{6}$
50	281·0	518	7·00	1·945	$\frac{1}{7}$
55	287·1	474	8·00	2·079	$\frac{1}{8}$
60	292·7	437	9·00	2·197	$\frac{1}{9}$
65	298·0	405	10·00	2·302	$\frac{1}{10}$
70	302·9	378			
75	307·5	353			
80	312·0	333			
85	316·1	314			
90	320·3	298			
95	324·1	283			
100	327·9	270			

The above table of Hyperbolic Logarithms is given for the convenience of calculating the average pressure for a given initial pressure and ratio of expansion.

R = ratio of expansion.
I = initial pressure.
a = average pressure throughout the stroke.

$$a = \frac{(1 + Hyp.\ Log.\ R)\ I}{R}.$$

CONDENSING WATER.

W = weight of steam to be condensed.
T = its temperature.
l = its latent heat.
W' = weight of injection water required.
t = its temperature.
t' = temperature of the mixture.

$$W' = \frac{(T + l - t')\ W}{t' - t}.$$

For values of l, see table on page 33.

Pressure of Steam in lb.	Latent Heat of Steam.	Sum of Latent and Sensible Heat.	Tempera-ture—Fahr.	Pressure of Steam in lb.	Latent Heat of Steam.	Sum of Latent and Sensible Heat.	Tempera-ture—Fahr.
	1092·6	1124·6	32°	29	939·6	1187·6	248°
	1080·0	1130·0	50°	39	927·0	1193·0	266°
	1067·4	1135·4	68°	53	914·4	1198·4	284°
	1054·8	1140·8	86°	69	901·8	1203·8	302°
1	1042·2	1146·2	104°	20	889·2	1209·2	320°
2	1029·6	1151·6	120°	115	874·8	1212·8	338°
3	1017·0	1157·0	140°	146	862·2	1218·2	356°
4·5	1004·4	1162·4	158°	182	849·6	1223·6	374°
7	991·8	1167·8	176°	229	835·2	1227·2	392°
10	979·2	1173·2	194°	279	822·6	1232·6	410°
*14·706	966·6	1178·6	†212°	348	808·2	1236·2	428°
21	952·2	1182·2	230°	402	795·6	1241·6	446°

* Atmospheric pressure.   † Boiling point.

WEIGHT OF BOILERS.

Reckon 6 lb. per $\frac{1}{8}''$ of thickness of plate for each superficial foot. This will give the weight very near, including lap and rivets.

STRENGTH OF BOILERS.

Strength of plate = 100.
Strength of single riveted joints = 56.
Strength of double riveted joints = 70.

STRENGTH OF BOILER TUBES.

P = collapsing pressure in lb.
K = thickness of plate in inches.
L = length of tube in feet.
D = diameter of tube in inches.

$$P = 806300 \frac{K \times 2·19}{L \times D}; \text{ or}$$

$Log. \ P = 1·5265 + 2·19 \ log. \ 100 \ K - log. \ (L \ D).$

CHIMNEYS.

F = number of lb. of coal consumed per hour.
h = height of chimney in feet.

D

A = area of chimney at top in square inches.

$H.\text{-}P.$ = indicated horse-power.

$$A = \frac{15\,F}{\sqrt{h}};$$

for economical pumping engines—

$$A = \frac{70\,H.\text{-}P.}{\sqrt{h}}.$$

Horse-power.	Height of Chimney.	Diameter at Base.	Diameter of Flue.	
	feet.	feet.	feet.	inches.
50	75	8	2	4
100	100	11	3	0
200	120	13	4	0
300	160	14	4	6

VELOCITY OF ARTIFICIAL DRAUGHT.

H = height of chimney in feet.

T = temperature of air supplying the chimney.

t = temperature of air at top of chimney.

V = velocity in feet per second.

$$V = 36 \cdot 5 \sqrt{H(T - t)}.$$

SHORT LINK CHAINS—BEST TESTED.

Size.	Working Load.	Outside Dimensions.
inch.	tons.	inches.
$\frac{3}{8}$	1	$1\frac{7}{8} \times 1\frac{1}{4}$
$\frac{7}{16}$	$1\frac{1}{2}$	$2\frac{1}{16} \times 1\frac{7}{16}$
$\frac{1}{2}$	2	$2\frac{1}{4} \times 1\frac{5}{8}$
$\frac{9}{16}$	3	$2\frac{7}{16} \times 1\frac{7}{8}$
$\frac{5}{8}$	4	$2\frac{3}{4} \times 2\frac{1}{16}$
$\frac{11}{16}$	5	$3\frac{1}{16} \times 2\frac{3}{8}$
$\frac{3}{4}$	6	$3\frac{3}{8} \times 2\frac{1}{2}$
$\frac{13}{16}$	7	$3\frac{3}{4} \times 2\frac{13}{16}$
$\frac{7}{8}$	8	$4\frac{1}{16} \times 3$
$\frac{15}{16}$	9	$4\frac{3}{8} \times 3\frac{3}{16}$
1	10	$4\frac{5}{8} \times 3\frac{1}{2}$
$1\frac{1}{16}$	$10\frac{1}{2}$	$4\frac{7}{8} \times 3\frac{3}{4}$
$1\frac{1}{8}$	12	$5\frac{3}{16} \times 4\frac{1}{16}$

TABLE OF OBSERVATIONS UPON TEN CORNISH ENGINES. BY JOHN S. ENYS, ESQ., 1838.

			CONSOLIDATED MINES.						UNITED MINES.			
			Taylor's Engine.	Davey's Engine.[1]	Job's Engine.[2]	Woolf's Engine.[3]	Bawden's Engine.	Pearce's Engine.[4]	Cardozo's Engine.[5]	Eldon's Engine.[6]	Loam's Engine.[7]	Hocking's Engine.
DIMENSIONS OF ENGINE.	Diameter of Cylinder	in inches	85	80	65	90	90	65	90	30	85	85
	Stroke of Piston	in feet	10	11⅓	9	10	10	9	9	9	10	10
	Diameter of Steam Valve	in inches	12	13	9	8	8	7	10	5	10	12
	Ditto of Equilibrium Valve	"	16	18	12	16	16	12	13	7	16	16
	Ditto of Exhaust Valve	"	20	24	14	19	19	14	15	10	19	19
DIMENSIONS OF BOILERS.	Number of Boilers		4	3	2	4	3	3	3	1	3	3
	Length of Boilers	in feet	{3·36 / 1·40}	37	{31 / 32}	35	36	36	36	36	{1·32 / 2·38}	44
	Diameter of Boilers	"	6½	7	6¼	6¼	6¼	6¼	6¾	6½	6½	6½
	Ditto of Tubes	"	{3· 3¾ / 1· 4}	4½	3¾	3¾	3¾	3¾	3¾	4	4	4
	Length of Fire Bars	"	4	4	4	4	4	4	4	4	4	4
	Total Area of Fire Bars	"	63	52	30	60	45	45	48	16	18	48
	Heating Surface Exposed to Flues	"	3781	3151	1598	3481	2694	2694	2694	941	2952	3451
	Water Space	in cubic feet	2467	2025	1033	2140	1650	1650	1650	579	1706	2085
	Steam Space	"	735	580	315	608	468	468	468	178	528	645
TEMPERATURES OBSERVED.	Open Air	Fahr.	57°	57°	—	57°	57°	51°	55°	56°	51°	55°
	Engine House	"	—	—	—	—	—	67°	—	65°	63°	66°
	Ashes over Boiler	"	80°	80°	81°	98°	98°	97°	88°	99°	79°	82°
	Cylinder Cover	"	111°	90°	102°	126°	96°	109°	94°	98°	90°	84°
	Middle of Cylinder Clothing *	"	W 77°	W 76°	B 95°	B 102°	B 140°	—	B 79°	W 60°	W 67°	W 68°
	Clothing of Steam Pipe	"	79°	—	—	130°	95°	—	140°	97°	80°	82°
	Condensing Water	"	64°	84°	58°	115°	110°	57°	60°	61°	63°	60°
	Hot Well	"	98°	100°	91°	140°	140°	97°	104°	94°	102°	96°
Height of Condenser Barometer		in inches	27¾	27¼	—	25⅔	—	—	—	—	—	27½
Number of Plunger Pumps			9	12	2	7	8	9	8	1	5	5
Number of Bucket Pumps			2	2	—	1	2	2	2	—	4	3
Water Load per square inch of Piston		in lb.	11·46	13·12	8·78	11·56	8·3	16·8	11·5	17·96	11·95	13·58
Proportion of Stroke where Steam cut off			1/3	1/3	—	1/3	1/3	1/4	1/3	1/3	2/11	—
Strokes per minute			8½	7½	—	9	9	10	8¼	9	8	7
Proportion of duration of in-door to out-door Stroke			4 : 7	5 : 8	4 : 7	5 : 7	5 : 7	5 : 5	4 : 6½	4 : 7	1 : 2	4 : 7½
Grease used per day		in lb.	12	12	10	12	12	10	12	6	12	12
Oil used per day		in pints	1	1	1	1	1	1	1	½	1	1
Men Employed			4	4	3	4	4	3	4	3	4	4
Boys Employed			3	3	3	4	4	3	3	—	3	3

[1] Six years old. The Show Engine of the Mine. [2] Employed in raising water forty-seven fathoms for the wheels and injection. [3] Cylinder in bad condition; since changed. [4] Old Engine; 1820. [5] Old Engine; no steam jacket. [6] Old Engine refitted. [7] An old 90-inch Cylinder used for the steam jacket of this engine.
* The prefix of the letter B signifies that the casting is of Brick; and W of Wood.

WEIGHT OF WROUGHT IRON.

The weight of 1 cubic inch of wrought iron = 0·28 lb.

$$\frac{\text{Number of cubic inches}}{8000} = \text{tons.}$$

$$\frac{\text{Number of cubic inches}}{400} = \text{cwts.}$$

$$\frac{\text{Number of cubic inches}}{100} = \text{qrs.}$$

$$\frac{\text{Number of cubic inches}}{3·5} = \text{lbs.}$$

WEIGHT OF CAST IRON.

The weight of 1 cubic inch of cast iron = 0·26 lb.

$$\frac{\text{Number of cubic inches}}{8640} = \text{tons.}$$

$$\frac{\text{Number of cubic inches}}{432} = \text{cwts.}$$

$$\frac{\text{Number of cubic inches}}{108} = \text{qrs.}$$

$$\frac{\text{Number of cubic inches}}{3·85} = \text{lbs.}$$

1 cubic foot of cast iron = 448 lb. = $\frac{1}{5}$th ton = 4 cwt.

1 foot superficial 1 inch thick = $\dfrac{1 \text{ cwt.}}{3}$.

Let n = number of cubic inches of cast iron.

W = weight in lb.

$W = \dfrac{n}{4} +$ (4 × the number of hundreds expressed by the first, or left hand figure in product), thus: Let n = 1728 cubic inches, then $= \dfrac{1728}{4} = 432$ and $4 \times 4 = 16 \therefore W = 432 + 16 = 448$ lb.

HORIZONTAL ENGINE AND DOUBLE RAM PUMPS. 580 H. POWER.

This Engine works a Pair of 20' Plunger Pumps
10 f.t Stroke. forcing 700 f.t in. One lift.

HENRY DAVEY
ENGINEER.

HORIZONTAL ENGINE AS APPLIED BETWEEN TWO PITS.

HENRY DAVEY
ENGINEER.

HORIZONTAL ENGINE WITH SINGLE QUADRANT.

HENRY DAVEY
ENGINEER.

Plate 4.

VERTICAL ENGINE AS APPLIED IN WATERWORKS &c.

Type Nº 4.

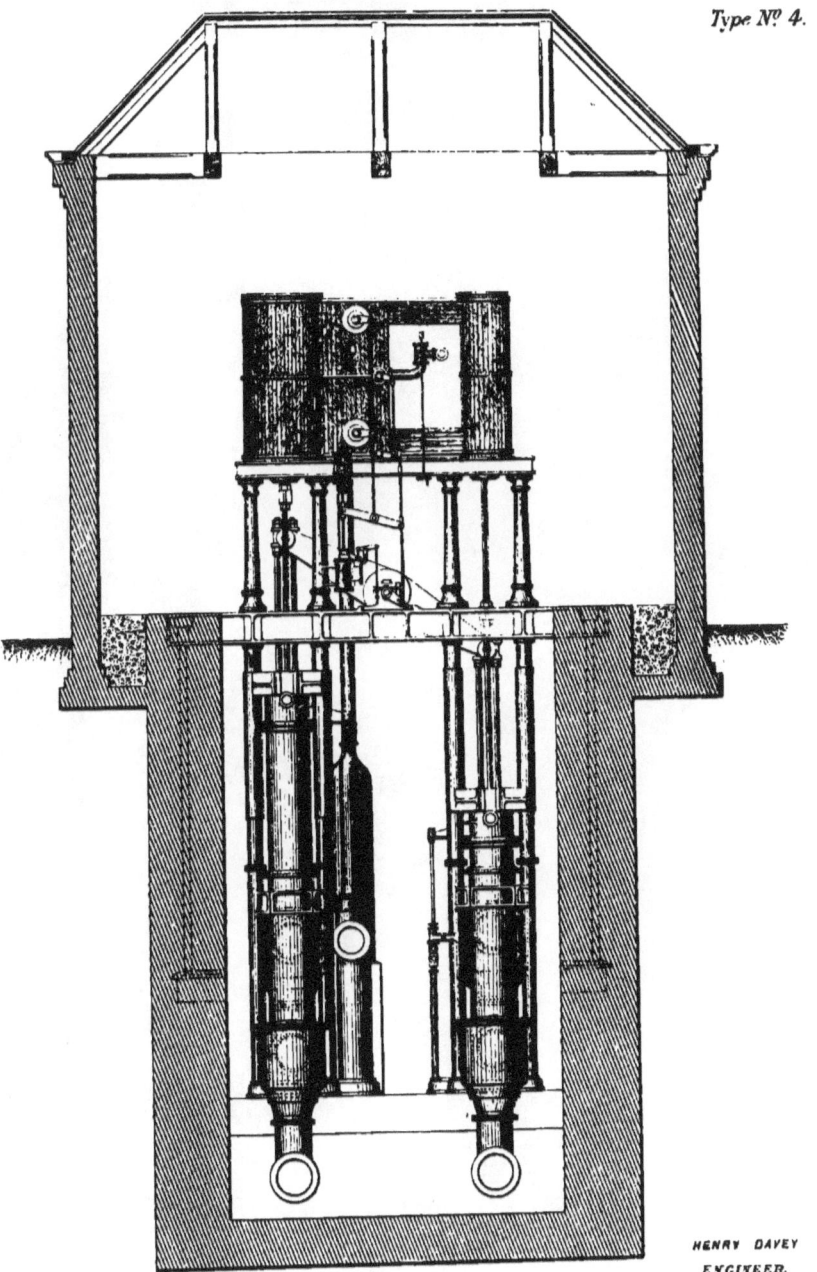

HENRY DAVEY
ENGINEER

Plate 5.

VERTICAL DEEP WELL ENGINE AS APPLIED IN WATERWORKS.

Type N? 5.

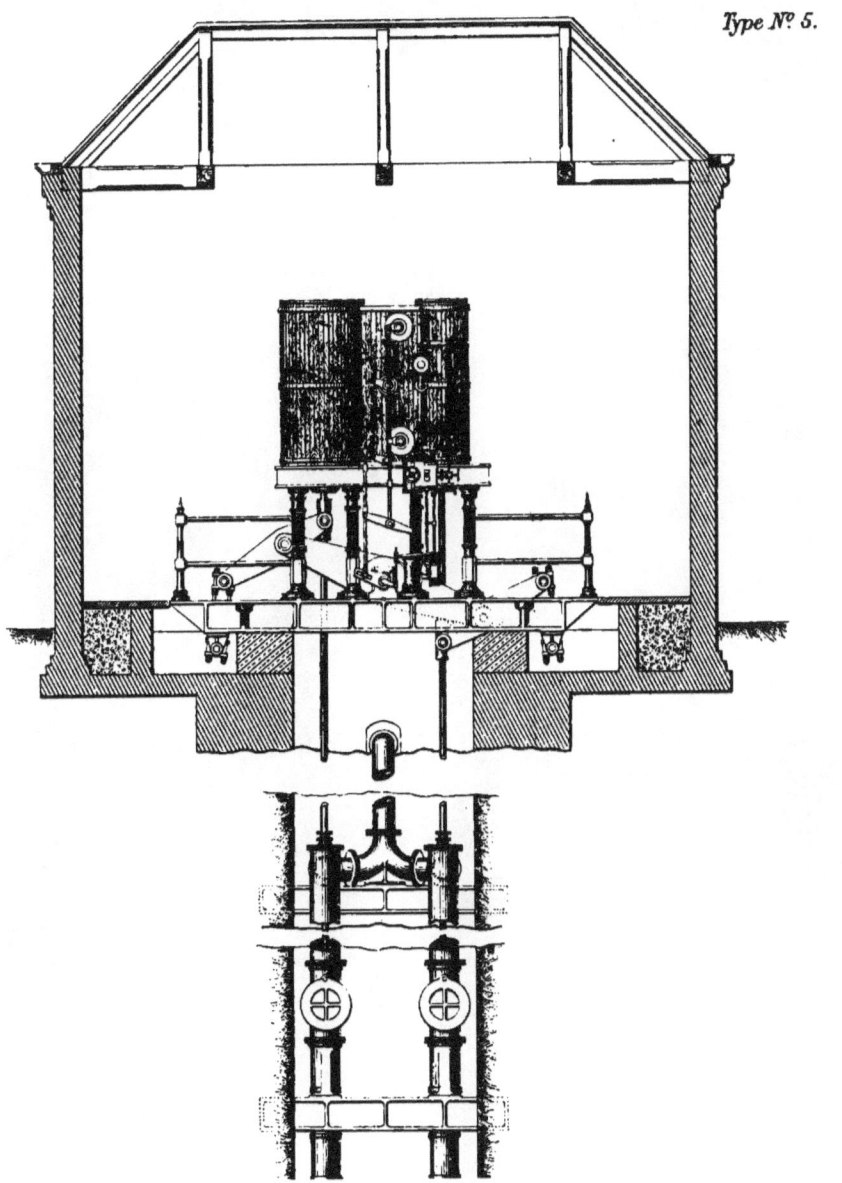

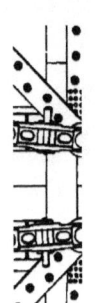

Plate 7.

Type No 7.

PUMPING ENGINES.

Fig. 16.

COMPOUND DIFFERENTIAL ENGINE
AND HYDRAULIC ENGINE
APPLIED UNDERGROUND.

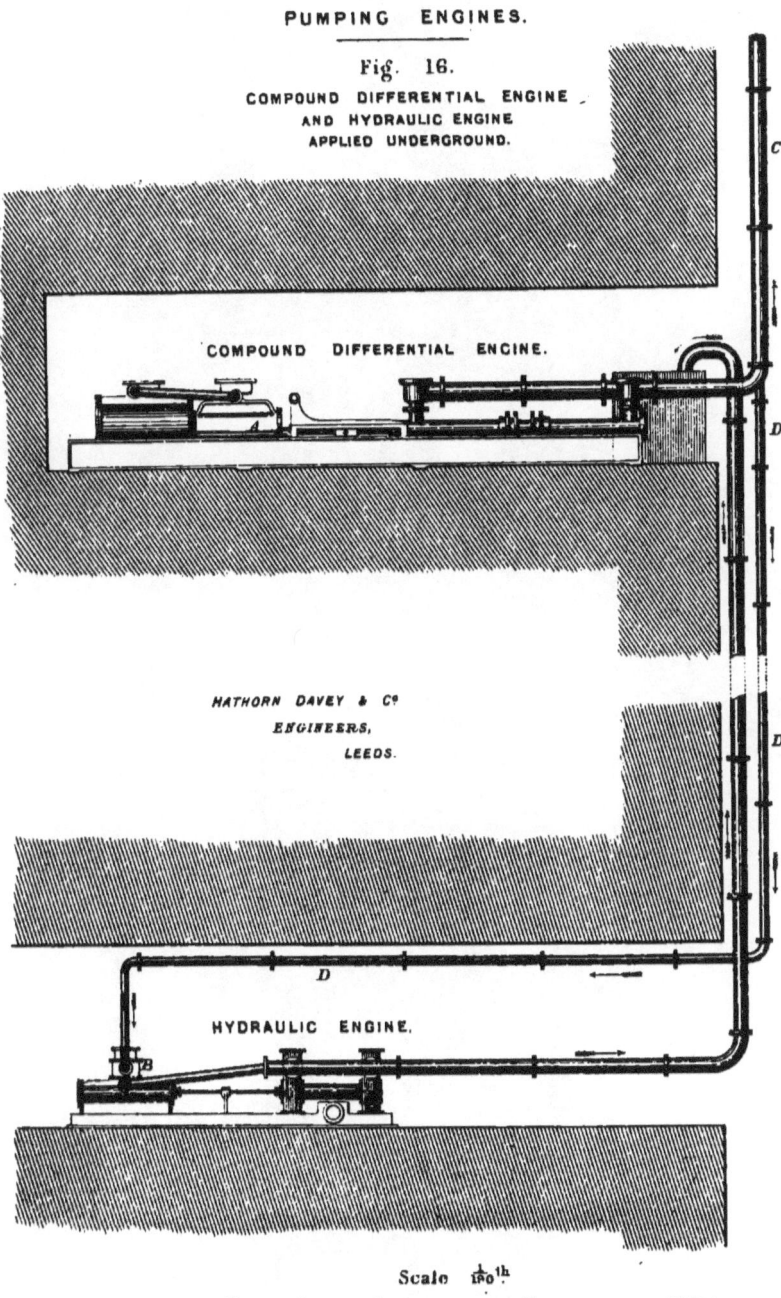

COMPOUND DIFFERENTIAL ENGINE.

HATHORN DAVEY & Cᵒ
ENGINEERS,
LEEDS.

HYDRAULIC ENGINE.

Scale ¹⁄₁₈₀ᵗʰ

10   5   0   10   20 Feet

HENRY DAVEY

Plate 8

IMPROVED PLUNGER PUMPS

Pumps 20" dia & 10 ft Stroke
lift 720 feet.

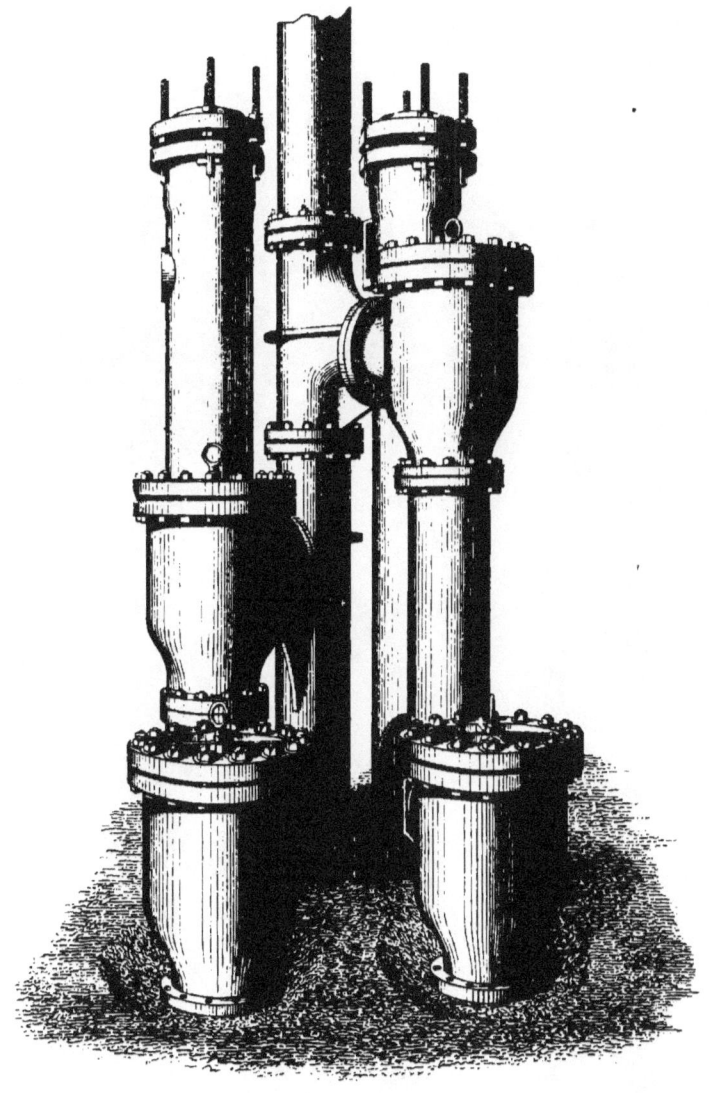

HENRY DAVEY
ENGINEER.

COMPOUND DIFFERENTIAL ENGINES.

Diagrams
Illustrating Action of Differential Valve Gear.

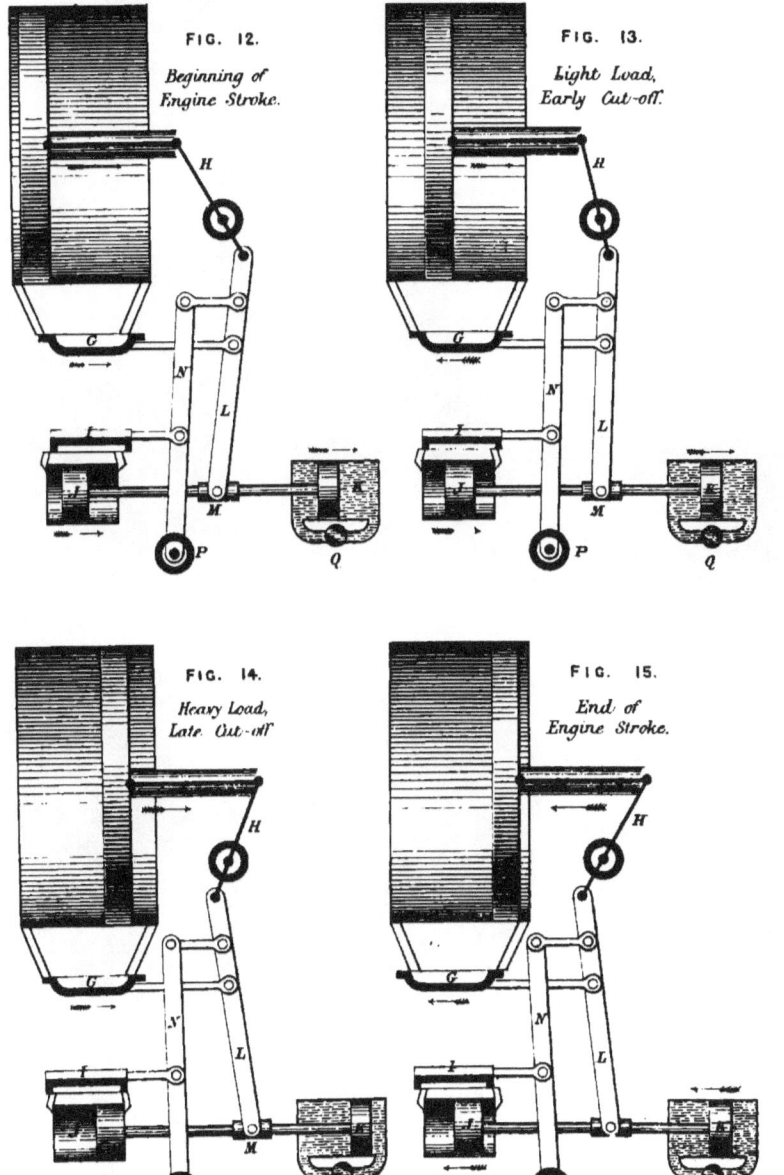

FIG. 12.

*Beginning of
Engine Stroke.*

FIG. 13.

*Light Load,
Early Cut-off.*

FIG. 14.

*Heavy Load,
Late Cut-off.*

FIG. 15.

*End of
Engine Stroke.*

HENRY DAVEY
ENGINEER.

COMPOUND DIFFERENTIAL ENGINE.

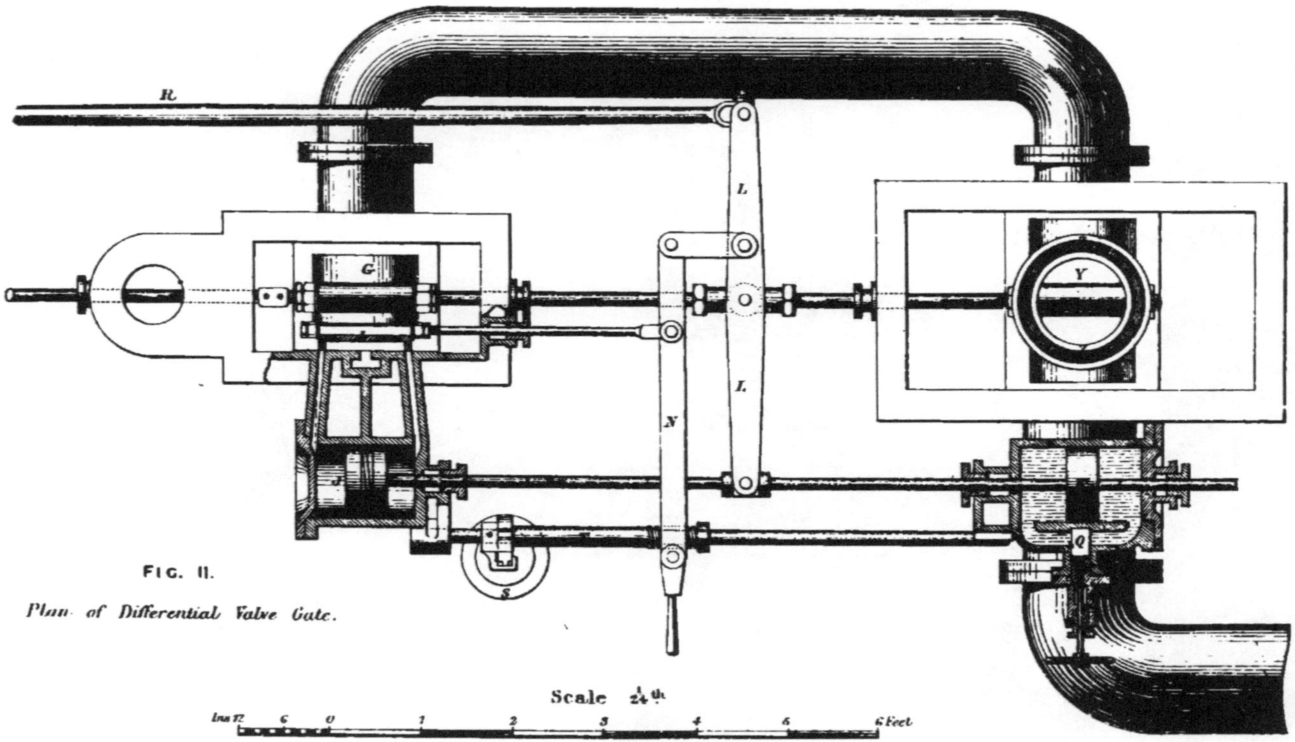

HENRY DAVEY
ENGINEER

FIG. II.

Plan of Differential Valve Gate.

Scale 2¼ᵗʰ